# Maple for the Calculus Student: A Tutorial

# Maple for the Calculus Student: A Tutorial

**Wade Ellis, Jr.**
West Valley College

**Ed Lodi**
West Valley College

**Brooks/Cole Publishing Company**
**Pacific Grove, California**

**Brooks/Cole Publishing Company**
A Division of Wadsworth, Inc.

ⓒ1989 by Wadsworth, Inc., Belmont, California 94002. All rights reserved. No part of this book may be reproduced, stored in a retrieval system, or transcribed, in any form or by any means—electronic, mechanical, photocopying, recording, or otherwise—without the prior written permission of the publisher, Brooks/Cole Publishing Company, Pacific Grove, California 93950, a division of Wadsworth, Inc. Maple is a registered trademark of the University of Waterloo. The Maple logo is a registered trademark of the Symbolic Computation Group, University of Waterloo. Macintosh is a registered trademark licensed to Apple Computer, Inc.

Printed in the United States of America
10  9  8  7  6  5  4  3  2  1

**Library of Congress Cataloging-in-Publication Data**
Ellis, Wade.
    Maple for the calculus student / Wade Ellis, Jr., Ed Lodi.
      p.   cm.
    Includes index.
    ISBN 0-534-11874-7
      1. Maple (Computer program)   2. Algebra—Data processing.
3. Calculus—Data processing.   I. Lodi, Ed.   II. Title.
QA155.7.E4E44   1989
512'.0285'5369—dc20                      89-31308
                                             CIP

Sponsoring Editor: *Robert Evans, Jeremy Hayhurst*
Editoral Assistant: *Jennifer R. Greenwood*
Production Editor: *Nancy L. Shammas, Ben Greensfelder*
Manuscript Editor: *Linda Thompson*
Permissions Editor: *Carline Haga*
Interior Design: *Katherine Minerva*
Cover Design: *Lori Hughes Design*
Compositor: *the Computer Tutors of San Jose*
Cover Printing: *Malloy Lithographing, Inc., Ann Arbor, Michigan*
Printing and Binding: *Malloy Lithographing, Inc., Ann Arbor, Michigan*

Customer support is available to registered users of Maple between 9 a.m. and 4 p.m. Pacific Time at (408) 373-0728, or E-mail via AppleLink D2248.

# Contents

# 8 Differential Equations 55

# 9 Linear Algebra 59

# Index 65

# Introduction

### Learning Maple

Maple is an extremely powerful computer-based tool for solving problems. Computer algebra systems like Maple are used by engineers, scientists, and mathematicians to assist them in attacking problems that require complicated and involved symbolic or numeric computations. As with hand-held calculators, these systems were expensive and cumbersome in their original form. Now almost every school-age child has a calculator and all scientists have access to computer-based computational tools, usually on their desks. Some problems can best be approached with paper and pencil, others with a simple calculator. Still other more complicated problems may require the use of sophisticated tools such as programmable calculators or computer-based software packages. As you develop your problem-solving skills, you should also be developing a feel for the tools that are appropriate for the problem at hand. In learning to use Maple, you will develop a valuable skill and a powerful asset in attacking problems throughout your career as a student or professional.

### Origins of Computer Algebra Systems

Computer algebra systems like Maple were first developed in the late 1950s by computer scientists interested in artificial intelligence. Although many believe that mathematics is computational in nature, it actually is the study of logical approaches to problems—logical rules are developed to use in solving problems. The rules (and their development) are the better part of mathematics; computation with these rules, though difficult, is not the essence of mathematical thought. Computers (with an enormous helping hand from human computer programmers) are much better at rule-based computation than mere mortals. Computer scientists have had much more difficulty programming computers to discern useful and logically correct rules or theorems. Maple, as an up-to-date computer algebra system, is an elegant tool for symbolic manipulations that can be of great help in solving problems, but it is not a substitute for analytic thought.

## Maple Capabilities

Maple is able to perform the following kinds of activities. It can symbolically factor polynomials; provide exact solutions to polynomial, exponential, trigonometric, and logarithmic equations; graph various functions; and perform operations on matrices with variables as entries. In addition, Maple can perform many operations that arise in calculus, including differentiation, integration, and the taking of limits. Performing these myriad operations have in past centuries been thought to require high intellect. That a computer can perform them seems to give the computer intelligence, when in reality the intelligence exists in the programming done by human beings.

You can find exact solutions to many difficult problems using Maple's operators or functions. Maple can also provide approximate solutions to many mathematical problems for which there are no algorithms for exact solutions. You can use Maple to check the results that it provides, and you are encouraged to do so, since very complicated programs are unlikely to be error free.

Maple's graphing capabilities are especially useful in attacking problems that occur in calculus and physics. Graphs can give you additional insight into scientific functions and expressions by presenting a visual as well as symbolic representation.

## The Maple Macintosh Environment

The Macintosh environment for Maple is especially easy to learn and use. If you are unfamiliar with the Macintosh environment, you should go through the *Guided Tour* that comes with your computer. The Macintosh popdown menus allow you to easily edit and change commands that you have already entered. The Macintosh windowing capability allows you to view several different representations of the same problem available at the same time. The excellent graphics screen is restful on the eyes and gives you a crisp display of Maple graphs.

## Purpose of the Book

This book is intended to provide you with a sound foundation for using the enormous power of Maple. Although the tutorial itself introduces just a few of the Maple operators or functions, it does present a representative sample of how such functions are entered and used. We have attempted to provide you not only with the basic syntax of Maple statements but also with a feel for how several Maple statements can be used together to investigate a particular problem. When you have completed the tutorial, you should be able to think of ways to use suites of Maple statements to provide insight into other problems you may encounter.

## How to Use This Book

You can obtain the greatest benefit from this tutorial by thoughtfully working through each item. You should carefully consider the statements you are asked to enter and think about the expected result before you enter the statement. Occasionally, you will be asked to enter statements that will give rise to error messages. This is to provide experience with such errors and to encourage you to experiment fearlessly with Maple statements.

# 1 The Maple Environment

Turn on your Macintosh computer. Use the mouse to open the Maple folder (if necessary) and the Maple 4.2 application. Your screen should look like this:

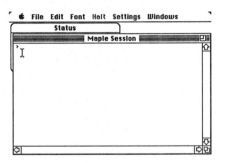

The > symbol with the blinking vertical line (|) on the next line is the standard Maple invitation to enter statements.

## Maple and Mathematics

You can easily learn to use the powerful capabilities of Maple if you thoughtfully work through the following tutorial. It is important, therefore, for you to enter each of the Maple statements on the right. The sentences on the left give you a description of the operations on the right. Additional explanation follows the Maple statements that appear in **boldface**.

It is also important that you think about each of the Maple statements and the explanations that follow them. The intent of this presentation is to give you a feel for the interaction that occurs between you, the Macintosh computer, and Maple. After you are comfortable working with the Macintosh and Maple, you will work through examples that will help you learn how to investigate mathematical problems and concepts using Maple.

## Learning to Use Maple

*You can add two numbers.*

**2 + 3;**
Notice the semicolon (;) at the end of the line. It is used as a terminator for each statement in Maple. After you have typed the semicolon, you should press the key marked Enter at the lower right of the keyboard. You should not confuse the Enter key with the Return key. The answer 5 appears in the middle of the screen followed by a **>** sign at the left of the screen one line down. The blinking vertical line (|) appears on the next line indented two spaces.

*You can also add fractions.*

**2/3 + 1/7;**
The fractions are typed with a division sign (/). After you type the semicolon (;), you should press Enter. Notice that the answer, $\frac{17}{21}$, is displayed in the center of the screen (with a fraction bar and the 17 directly above the 21).

*A number can be raised to a power.*

**2^5;**
The caret (^) indicates exponentiation. You have asked Maple to find $2^5$. As you know, this is 32.

*What happens when you forget to type the semicolon (;) before you press Enter?*

**2 + 3**
To simulate this error, be sure not to type the semicolon (;) before you press Enter. The blinking vertical line (|) moves down two lines and nothing appears on the screen. You may wait for some time before you discover that no answer has appeared. A semicolon (;) is missing at the end of the statement. Thus, Maple is waiting for you to end the statement.

*You can correct this problem.*

**;**
Type a semicolon and press Enter. The answer 5 appears in the center of the screen when you do this. The Maple prompt (**>**) and the blinking vertical line (|) appear, inviting you to enter another statement.

*The following problem occurs frequently for beginners and experienced Maple users: pressing the Return key (the big key) when you should press the Enter key.*

**2 + 3;**
To see this error, be sure to press the Return key instead of the Enter key. The blinking vertical line (|) moves down to the first position on the next line. Nothing else happens on the screen. Pressing Enter is one way to recover from this problem. You should do that now. Maple returns to the standard entry environment.

## Numbers: Integers, Rationals, Decimals

*You can raise 2 to large powers.*

**2^32;**
Notice that the answer is displayed exactly. This result is different (and correctly so) from what you would obtain on a calculator.

*You can raise rational numbers to large powers as well.*

**(1/2)^32;**
The parentheses are for clarity. The fraction displayed is once again the exact answer.

*A different result occurs if you use another representation of one-half.*

`0.5^32;`

Notice that the displayed answer differs from the previous answer. How many digits are displayed? This answer is similar to what you would expect from a calculator and is the best 10-digit approximation to the exact answer.

*You can force the rational answer to be approximated by a decimal using the Maple* `evalf` *function.*

`evalf((1/2)^32);`

Take careful note of the parentheses. The outermost parentheses enclose the argument of the Maple function `evalf`. This result matches the result for the previous statement. It also has 10 digits.

*Mistakes often occur when you are entering expressions that have many parentheses in them.*

`evalf(1/2)^32;`

Be sure to enter the parentheses exactly as shown. The message displayed is a standard message that occurs when Maple does not understand what you entered. Here, there are more right parentheses than left parentheses. You should check for an incorrectly formed statement or a misspelled word whenever a syntax error message occurs.

*When the inevitable mistakes occur, you can recover.*

`evalf((1/2)^32);`

Notice the two left parentheses just after the `evalf` function name. If you left one of them out, you can correct the statement by retyping it. If you are familiar with Macintosh editing, you can copy the original statement, paste it just below the last **>** symbol, and insert the correction. Of course, you would press Enter to see the answer.

*You can specify the number of digits that will be displayed in decimal representations of numbers.*

`Digits := 20;`

You must type a capital D in Digits since Maple differentiates uppercase and lowercase letters in names. The := symbol (with no space between the : and =) instructs Maple to assign the value on the right to the variable on the left. This increases the number of digits displayed from the standard 10 to 20. You see that the variable *Digits* is now 20 by the displayed result:

```
Digits := 20.
```

*Let's look at the decimal representation of $2^{100}$.*

```
evalf(2^100);
```

Here the result is displayed in 20 digit decimal format. This is also called floating-point format.

## Variables

*You can assign a value to a variable.*

```
x := 5;
```

This statement assigns the value 5 to the variable $x$. In previous interactions with Maple, you saw an answer. Here, Maple echoes the statement you entered (without the ending semicolon). Maple echoes the statement you enter when no operation is performed that results in an answer.

*Maple can evaluate expressions.*

```
x^2;
```

The variable $x$ has the value 5. Maple assigns $x$ this value and then evaluates the expression.

*Multiplication must be explicitly indicated in the expression to be evaluated.*

```
2*x;
```

Multiplication is indicated by an asterisk. The value of the expression when $x$ is 5 is clearly 10.

*What happens if you forget an asterisk (\*)?*

```
2x;
```

The syntax error message indicates that there is a problem with the statement as entered. You have to determine what correction(s) must be made. Of course, you must reenter the statement with the multiplication symbol correctly placed.

*You can reestablish x as a variable with no value assigned to it.*

```
x := 'x';
```

The ′ symbol is on the key to the left of the Return key. You might wish to think of this as assigning the value $x$ to the variable $x$.

*You can check that $x$ has no assigned value.*

```
x^2 + 4*x;
```
The expression you entered is displayed in a different way. Notice that $x$ squared is displayed in standard mathematical notation. Also, there is no asterisk in the displayed expression, but there is a space between 4 and $x$. While you *must* enter expressions with the exponentiation and multiplication symbols, Maple displays expressions without them.

*The " symbol is assigned the value of the last expression displayed.*

```
";
```
The expression $x^2 + 4x$ is displayed in Maple format.

*You can use the " in expressions.*

```
" - 2*x;
```
Maple combines like terms when possible, as you can see.

*You can change the expression with other algebraic operations.*

```
" - 3;
```
Here, the $-3$ is simply appended to the expression. Of course, this means that 3 is subtracted from the expression $x^2 + 2x$.

*There is a* factor *function in Maple.*

```
factor(");
```
The factors of the expression assigned to the " symbol are displayed.

*You can check this result using the* expand *function.*

```
expand(");
```
You should attempt to check your results. Maple is a large and complex program that has errors (bugs) in it, as all large programs do. Sometimes Maple (and hand calculations) are wrong. You should develop the habit of checking your Maple computations in the same way you have developed habits and methods for checking your hand computations.

*An expression can be assigned to a variable name. Here you will use the " symbol in the assignment statement.*

```
p := ";
```
Note that $p$ is assigned the value $x^2 + 2x - 3$.

*You can always check what value is assigned to a variable.*

**p;**
The value displayed is the value of *p*.

*You can factor the expression p.*

**factor(p);**
You should check to see that the factors of *p* are the same as the factors of $x^2 + 2x - 3$ given earlier.

*Has factoring p changed p?*

**p;**
The variable *p* has remained the same.

*You can solve the equation $x^2 + 2x - 3 = 0$ or $p = 0$.*

**solve(p=0);**
The solution is a set of two integers. These integers, as you know, are the solutions to the two equations $(x+3) = 0$ and $(x - 1) = 0$.

*You can plot or graph an expression. You can indicate the domain of the variable x in the expression p.*

**plot(p, x = -4..4);**
Notice how the domain is specified. You *must* use the two periods rather than a comma. The graph will be displayed after a short time in a separate window.

   Does the graph intersect the *x*-axis at the same values you obtained solving the equation $p = 0$? You can click on the Close box in the upper left corner of the Plot window in order to enter further Maple statements.

*You can, if you wish, specify the range values.*

**plot(p, x = -4..4, y - -10..10);**
Be sure there are two periods in the range specification. Notice that the tick marks on both the *x*− and *y*−axes are integers. Also, this second Plot window is slightly lower on the screen. Click on the Close box.

*There is a common mistake made in entering the* plot *function.*

**plot(p, x = -2,3);**
The error message clearly indicates where the problem occurs. When you see this message, you should first check to be sure you have entered the domain and range with periods.

*You can easily change the domain of your graph.*

```
plot(p, x = -2..3, y = -10..10);
```
You can easily change the domain and range of your graphs to explore the behavior of an expression in detail. You can move the Plot window up by clicking and dragging its Title bar.

## Editing

You can use the powerful editing features of the Macintosh to speed your work with Maple.

*You can copy the last* plot *statement.*

Use the mouse to move the blinking cursor to the beginning of the previous plot statement. Hold down the mouse button, drag the cursor to the end of the statement, and release the mouse button. You will have performed this action correctly if the entire plot statement is highlighted in black. Now pull down the Edit menu by pointing at it and holding down the mouse button. Then select Copy by continuing to hold down the mouse button and dragging the highlight down to Copy. Release the mouse button. An image of the plot statement is now saved.

*You can paste this copied statement anywhere you wish in the Maple environment.*

Move the blinking cursor below the last **>** symbol and click the mouse button. Pull down the Edit menu and select Paste. The plot statement should appear at the blinking cursor position.

*You can edit this statement so that the domain of the plot is from* −3 *to* 3.

To do this, move the blinking cursor to the right of −2 in the plot statement. Now press the Delete or Backspace key to erase the 2. Type 3 and press Enter. Notice that you did not have to be at the end of the line for Maple to accept the entire statement when you pressed Enter.

If you pull down the Edit menu, you can see that there are special key combinations that can be used to accomplish copying and pasting. For example, if you highlight a word, you can copy it by holding down the key with the spider on it and pressing C. You can speed your work by using these key combinations. The Undo command in the Edit menu is very useful if you are surprised by an edit operation. You can use the keys with arrows on them on your keyboard to move the cursor around the screen.

## Printing and Saving

You can print a selected window or the screen using the Print Window or Print Screen command on the File menu.

It is possible to save a series of Maple statements by opening a new window using the New File command of the File menu and entering the statements there. The New File command creates a window called the Text Window that can be saved using the Save command from the File menu. When statements in a Text Window are entered, the results appear in the Maple Session Window.

## Summary

### Operations

The standard mathematical operations of addition, subtraction, multiplication, division, and exponentiation are available in Maple. The addition and subtraction symbols are the usual ones. Multiplication is indicated with an $*$, division with a $/$, and exponentiation with the $\wedge$ when entering arithmetic or algebraic expressions. Multiplication is indicated by a space, whereas exponentiation is indicated in standard mathematical notation in expressions displayed by Maple.

## Types of Numbers

Maple recognizes integers, rational numbers, and floating point numbers (decimals). Integer and rational operations are done exactly. Floating-point computations are approximations of the actual values. Sometimes these floating-point approximations are exact, but frequently they are not. You can change the number of digits displayed and used in computations by assigning values to the *Digits* variable. You should exercise care in using floating point numbers.

When you wish to, you can convert exact rational numbers or integers to approximate floating-point notation using the `evalf` function.

## Functions

Maple is endowed with a multitude of interesting and powerful functions. The `solve`, `factor`, `expand`, and `plot` functions are but a few of these functions. You used these functions together to explore the polynomial expression $x^2 + 2x - 3$.

## Additional Activities

### Entering Expressions

Write each of the following expressions as you would enter them in Maple:

1. $\dfrac{1}{x-2}$

2. $\dfrac{1}{x} + \dfrac{5}{3x}$

3. $x - 2\,x^4 - 3x^3 + 1$

4. $2^x$

5. $2^{x+5}$

## Exploring Functions

You may wish to explore the following expressions using the `solve`, `factor`, and `plot` functions.

1. $x^2 - 5x + 6$
2. $x^2 - 4x - 12$
3. $6x^2 + x - 15$
4. $40x^2 - 131x + 84$
5. $90x^2 - 249x + 168$

You've done a lot so far. Perhaps you would like to take a break and come back later to continue the tutorial.

# 2   Solving Equations

## Polynomial Equations

You've seen how to solve some equations. Let's look at some more.

```
solve(3*x^2 - 5*x + 2 = 0);
```
This is $3x^2 - 5x + 2 = 0$. Use your factoring ability to check the solutions displayed.

Sometimes solutions are irrational numbers.

```
solve(x^2 - 3 = 0);
```
Maple uses fractional exponents rather than radicals to display irrational numbers and expressions

You can solve equations whose solutions are more complicated expressions.

```
solve(3*x^2 - 5*x + 1 = 0);
```
The two displayed solutions are, as before, separated by a comma. Again, radicals are represented by fractional exponents.

You may wish to have decimal approximations for these solutions.

```
fsolve(3*x^2 - 5*x + 1 = 0);
```
The number of digits in the decimal approximations will depend upon the value of *Digits*.

Some equations have complex solutions.

```
solve(x^2 + 1 = 0);
```
The complex number $i$ is represented as $I$. This may be difficult to read on the screen. You may wish to use the Font menu to change the font so that $i$ is easier to discern.

You can solve equations whose solutions are more involved complex expressions.

```
solve(3*x^2 - 5*x + 7 = 0);
```
If you have changed the font, you will more easily be able to read the complex solutions displayed. You need to be careful to look for $i$'s in solutions, since they can look like absolute value signs.

Again, you may wish to display the solutions in decimal form.

```
fsolve(3*x^2 - 5*x + 7 = 0);
```
The function `fsolve` finds only real number solutions.

The exact solutions can be approximated by decimals.

```
solve(3*x^2 - 5*x + 7 = 0);
```
This displays the exact complex solutions as a list or set as before.

You can assign this list to a variable.

```
s := ";
```
The " mark is the most recent displayed result.

The variable s is a list with two elements. You can access each solution separately.

```
evalf(s[1]);
```
Notice the brackets around 1. The function `evalf` returns the floating-point approximation of the first solution in the list.

You can access the second solution as well.

```
evalf(s[2]);
```
The variable s is a list. The second element of the list is designated as s[2] in Maple. Brackets (rather than parentheses) are required.

You can solve polynomial equations of higher degree than 2. You begin by assigning the polynomial expression to a variable.

```
q := 6*x^4 - 35*x^3 + 22*x^2 + 17*x - 10;
```
This is $6x^4 - 35x^3 + 22x^2 + 17x - 10$.

You can now solve the equation $q = 0$.

```
solve(q = 0);
```
You know that a polynomial of degree 4 will have at most four solutions. This equation has four rational solutions.

*A slightly different equation (q = 1) can give a much different result.*

```
solve(q = 1);
```
The solutions will go on for a very, very long time and are clearly very complicated. You may be able to see that portions of the solutions displayed have the complex number $i$ in them. The Halt Main menu item can be used to interrupt Maple only when it is not displaying information to the screen.

*You can get a much less complicated approximation to the solutions using the* `fsolve` *function.*

```
fsolve(q = 1);
```
The display of these solutions still take some time. Notice that the solutions are between $-1$ and 6 and are real numbers. This seems to contradict the results using `solve`. The $i$'s must cancel in the `solve` display.

*You can use* `plot` *to graph the expression $q - 1$ to resolve these conflicting results.*

```
plot(q - 1, x = -1..6);
```
The places where this graph crosses the $x$-axis are the solutions of the equation $q - 1 = 0$. The graph indicates that there are four real solutions that seem to correspond to the solutions obtained using `fsolve`. You can adjust the graphing window in the `plot` function to investigate this expression further. Here again, you have used the power of Maple to check the results you obtained with Maple. Click on the Close box before entering more Maple statements.

## Other Types of Equations

*You can solve trigonometric equations.*

```
p := cos(x) - sin(x);
```
Notice that parentheses are required in using the trigonometric functions.

*Now you can solve the equation $p = 0$, which is equivalent to $\cos(x) - \sin(x) = 0$.*

```
solve(p = 0);
```
You may recall that trigonometric equations often have an infinite number of solutions. The solution displayed on the screen, `1/4 Pi` or $\frac{\pi}{4}$, may not be the only solution. The `solve` function usually returns a single solution to nonpolynomial equations even if there are many solutions.

*Graphing the expression can assist you in determining the complete solution set.*

```
plot(p, x = 0..2*Pi);
```
The symbol $\pi$ is represented by Pi with a capital P. Notice that there is more than one solution.

*You can use the `fsolve` function to find other solutions.*

```
fsolve(p = 0, x, 1.5..4);
```
Notice the comma after x. The 1.5 . . 4 indicates which interval `fsolve` searches for a solution. The number displayed should be a multiple of $\pi$.

*What multiple of $\pi$ is it?*

```
evalf("/Pi);
```
Thus, the solution is approximately $\frac{5\pi}{4}$, which is $\pi$ units from the previous solution. You may wish to write the solution set in set notation based on this information.

*You can solve logarithmic equations also.*

```
solve(ln(x) + ln(x+1) = ln(2));
```
$\ln(x)$ is the natural logarithm of $x$. Thus, $\ln(2)$ is $\log_e(2)$ and is approximately 0.693172. The solutions to the equation are given as 1 and $-2$. The number $-2$ cannot be a solution, since $\ln(-2)$ is undefined. You may wish to solve this equation using paper and pencil methods to see why $-2$ arises as a solution. Logarithms to bases other than $e$ can be entered using the standard conversion formula. For example,

$$\log_{10}(x) = \frac{\ln(x)}{\ln(10)}$$

*Exponential equations are easy to enter and solve.*

```
solve(2^x = 5);
```
The exact solution is displayed using natural logarithms. You should use the `evalf` function to obtain a decimal approximation to the solution.

*You can use* `fsolve` *to find an approximate solution to this equation.*

`fsolve(2^x = 5);`
Notice that the solution is different from the solution you obtained using `evalf`. Maple uses two different methods to obtain this approximate result. The results differ in the tenth place. Better approximations can be obtained by setting *Digits* to a larger integer value.

## Inequalities

*You can solve inequalities.*

`solve(x^2 - 5*x < 0);`
The solution set for this inequality is the interval $(0, 5)$, or the set $\{x|0 < x < 5\}$. The solution is displayed by Maple as $\{x<5, \ 0<x\}$, a form of set notation meant to represent set intersection.

*Maple also uses a form of set notation meant to represent set union.*

`solve(x^2 - 5*x >= 0);`
Notice how the $\geq$ symbol is represented in Maple. Maple displays $\{x<=0\}$, $\{5<=x\}$. The solution uses two sets of braces to indicate the union of two sets as compared with the intersection notation which uses only one set of braces. You should be careful to look for these differences in notation when solving inequalities.

*You can also solve systems of equations.*

`solve({x+y=5,x-y=2}, {x,y});`
Here, Maple solves a set of equations (in the first pair of braces),

$$x + y = 5$$
$$x - y = 2$$

for a set of variables (in the second set of braces). The solution set is displayed with an $x$ value and a $y$ value.

*You can check the solution of the system of equations graphically.*

`plot({5-x, x-2},x=-4..4);`
Notice that you must write the equation $x + y = 5$ in the form $y = 5 - x$ and $x - y = 2$ in the form $y = x - 2$ to plot the graphs.

*A slight change in the equations gives a different result.*

```
solve({x+y=5.,x-y=2}, {x,y});
```

Notice the decimal point after the 5 in the first pair of braces. Maple uses a numerical method that yields approximate solutions if one of the numbers in the system of equations is in floating-point format. In this case, the approximations have the same value as the exact rational solutions displayed in the previous item.

## Summary

### The `solve` Function

The `solve` function gives exact solutions to equations. If the equation is a polynomial equation of degree 4 or less, then all the solutions (real and complex) are displayed. For polynomial equations of degree 5 or more or for nonpolynomial equations, the `solve` function will display those solutions that it finds. Maple uses the phrase RootsOf when it isolates an expression whose roots it cannot find. NULL is displayed if no solutions to an equation are found. You can also solve inequalities and systems of equations with this function.

### The `fsolve` Function

The `fsolve` function gives approximate floating-point solutions to equations. If the equation is a polynomial equation, then approximations are given for all real solutions. Usually, only one solution is given for nonpolynomial equations. You can specify an interval of values to use in searching for solutions. Particularly difficult (ill-conditioned) equations may cause `fsolve` to miss roots.

## Approximating Solutions Using `evalf`

The `solve` function will display some solutions as complicated expressions involving fractional exponents and complex numbers. You can approximate such complicated complex solutions by first assigning the set of solutions to a variable, say $s$. Then you can access each individual solution using a notation involving brackets. For example, the second solution in the list of displayed solutions would be $s[2]$. Finally, you use the `evalf` function to approximate the solution.

## Using the `plot` Function in Solving Equations

You can use the `plot` function to help you find locations (if any) where the graph crosses the horizontal axis. You can use these locations to determine intervals to use with the `fsolve` function. The graph can also resolve questions about complex and real roots of an equation. You may need to graph the function using different graphics windows to clearly see the behavior of the expression.

## Additional Activities

You may wish to explore the following polynomials using the `factor`, `solve`, `fsolve`, `evalf`, and `plot` functions.

1. $x^4 - x^3 - 5x^2 + 12$
2. $2x^3 - 13x^2 - 4x + 60$
3. $8x^2 + 2x^3 - x^4$
4. $2x^4 - 5x^3 + 10x - 12$
5. $x^5 - x^4 - 15x^3 + x^2 + 38x + 24$

# 3 Rational Expressions

You can assign a rational expression to a variable.

```
r := 1/(x + 1) - 1/(x - 1);
```
Since you must enter rational expressions without fraction bars, you need to be careful to delineate the numerators and denominators clearly using parentheses. Fortunately, Maple displays this rational expression in standard two-dimensional format.

You can simplify this expression.

```
simplify(r);
```
Maple adds the two rational expressions and displays the result in two-dimensional form with a leading $-2$.

You can extract the denominator from this expression using the denom function.

```
denom(simplify(r));
```
The denominator is displayed in factored form as it was displayed in the last item. You can use the expand function to display this denominator as a polynomial in standard form.

The numerator can be extracted with the numer function.

```
numer(simplify(r));
```
Maple considers the $-2$ to be the numerator of this rational expression.

You can graph this expression using the plot function.

```
plot(r);
```
Maple uses $-10$ to $10$ for $x$ values if no domain is specified. You know that this expression should have asymptotes from your experience in precalculus mathematics. However, the steep lines that appear to be asymptotes near $-1$ and $1$ are only lines joining points on the graph.

*You can force Maple to plot just the points that it computes for a graph of an expression.*

```
plot(r, style = POINT);
```
The POINT designation must be in capital letters. The graph is now displayed without connecting the computed points. There are no apparent asymptotic lines on the displayed graph.

*You can use the solve function to determine where the asymptotes should be on the graph.*

```
solve(denom(r));
```
The two solutions give the location of the asymptotes. They can be checked against the graph of the expression.

*You can graph the expression with a smaller domain that still includes the asymptotes.*

```
plot(r, x=-2..2, style = POINT);
```
Maple now displays a graph that more closely represents the known features of the graph of the expression.

*Let's look at another rational expression.*

```
s := (x^2 + 5*x + 6)/(x^3 + 2*x^2 - x - 2);
```
Notice the parentheses around the numerator and denominator of the expression

$$\frac{x^2 + 5x + 6}{x^3 + 2x^2 - x - 2}$$

*You can factor both the numerator and denominator of this rational expression.*

```
factor(numer(s));
factor(denom(s));
```
Here, you are entering two statements (pressing Enter after each statement). You can see that the numerator and denominator can be factored.

*You can simplify the expression s.*

```
simplify(s);
```
Now the original expression $s$ is seen to be in lowest terms. You should be aware that this simplification is correct only if $x \neq -2$.

*You can graph this expression.*

```
plot(s);
```
Notice that there is no vertical asymptote at $x = -2$, even though you saw that $x + 2$ was a factor of the denominator. There is no vertical asymptote at $x = -2$ because this factor also appears in the numerator. The Maple graph seems to be defined at $x = -2$, which is incorrect since the denominator is zero at $x = -2$. You should continue to be aware of possible discrepancies between Maple graphs (which are just connected points) and actual graphs.

## Summary

### Manipulating Rational Expressions

You use the `denom`, `numer`, `factor`, `solve`, and `simplify` functions to investigate and manipulate rational expressions—fractions whose numerators and denominators are polynomials. You can capture the numerator and denominator separately using the `numer` and `denom` functions. You can solve the numerator polynomial to find the zeros of the expression and the denominator to find the possible vertical asymptotes. You can use the `simplify` function to reduce a rational expression to lowest terms.

The `plot` function can be used to graph such expressions. You can determine the position of asymptotes in these graphs using the `solve` function or perhaps the `factor` function.

# Additional Activities

You may wish to explore the following rational expressions using the `factor` and `solve` functions:

1. $\dfrac{2x - 3}{x^2 - 9}$

2. $\dfrac{2x + 3}{x - 1}$

3. $\dfrac{x^2 - 2x - 8}{x^2 - 2x}$

4. $\dfrac{x^2 - 1}{x + 2}$

5. $\dfrac{2x^2 - 3x - 2}{x^2 - 5x}$

6. $\dfrac{4x^3 - 5x^2 + 3x - 6}{2x^2 + 3x + 5}$

# 4 Defining Functions

## One-Variable Functions

You can use the Maple procedure facility to create function definitions. A procedure performs a task that is described in a set of instructions. A function is a procedure that returns a value specified by its set of instructions. You can use this facility to define functions such as:

$$f(x) = x^2 + 3x - 5$$

$$g(x) = \cos(x) - x\ln(x)$$

$$h(x) = \frac{\cos(x)}{x^2 + 3x - 21}$$

*You can define the function f whose rule is $f(x) = x^2$.*

```
f := proc (x) x^2 end;
```
The procedure is assigned the name $f$. This procedure has one variable $x$ and one task (to square $x$). This procedure is a function because it returns exactly one value. Maple echoes the procedure definition when you enter this line. Notice that ** is used for exponentiation in echoed procedure definitions.

*You can obtain function values for f.*

```
f(2);
```
This statement asks Maple to evaluate and display the function value of $f$ at $x = 2$.

*You can use variable names in a function evaluation.*

```
f(a+b);
```
The value displayed is the value of $f$ at $x = (a+b)$. You can use the expand function to display this function value without parentheses.

*You can also use defined functions in algebraic expressions such as*

$$\frac{f(x+h) - f(x)}{h}$$

```
(f(x + h) - f(x))/h;
```
Parentheses must be placed around the entire numerator, since you are entering this expression on one line.

*The value displayed can be simplified.*

```
simplify(");
```
Maple removes the parentheses, combines like terms, and reduces fractions to lowest terms.

*You can also define functions that contain more than one rule, such as*
$$f(x) = \begin{cases} x^2 & \text{if } x > 3 \\ x - 5 & \text{if } x \leq 3 \end{cases}$$

```
f := proc (x) if x > 3 then x^2
else if x <= 3 then x - 5 fi fi end;
```
This single statement cannot fit on one line in this book. However, you should type it on a single line. In this procedure, if the domain value $x$ is greater than 3, then $f(x)$ is determined by the rule $x^2$. On the other hand (else), if $x$ is less than or equal to 3, then $f(x)$ is determined by the rule $x - 5$. The if phrase is terminated by fi (if spelled backwards). You must enter this statement carefully.

*You can obtain function values as before.*

```
f(2);
f(5);
```
Remember to press Enter after each semicolon. You should check to make sure the function is using the appropriate rule in each case.

*You can use* plot *to graph this function.*

```
plot(f);
```
The domain is automatically set to $[-10, 10]$. Does the graph of the function clearly show the two pieces of the graph? You should graph this function again using the style=POINT feature in plot to see a more accurate representation of the function.

## Functions of Several Variables

Functions of several variables occur in the latter part of the calculus sequence. They are written in mathematical notation much like one-variable functions:

$$f(x, y) = x^2 + y^2 - 3$$

is one example.

*You can define the preceding function in Maple.*

```
f := proc (x,y) x^2 + y^2 - 3 end;
```
Here, each domain value of the function is an ordered pair of numbers and the rule uses two variables.

*You can find values for such functions.*

```
f(2,5);
```
The function value for the pair of numbers (2, 5) is $2^2 + 5^2 - 3$. The number displayed is 26.

# 5

## Additional Features of the `plot` Function

The `plot` function has several options that allow you to create many types of graphs.

*Maple allows two or more functions to be graphed on the same axes.*

**`plot({x^2,2*x + 5});`**
You should watch carefully as the functions are drawn so that you can tell which graph is associated with $x^2$ and which is associated with $2x + 5$. In this case, you can easily tell the parabola from the straight line. You can change the graphing window if you wish to examine parts of these graphs (the intersections) more carefully.

## Parametric Form

You can graph curves expressed in parametric form. For example, the two equations

$$x(t) = t - 1 \text{ and } y(t) = t^2$$

give the $x$ and $y$ coordinates of a parabolic curve based on a parameter (dummy variable) $t$. If you solve these two functional equations for $y$ in terms of $x$ by eliminating the parameter $t$, you will find that $y = (x + 1)^2$, the equation of a parabola.

*The* `plot` *function is used with special grouping symbols, [ ], to graph the parametric curve just defined.*

`plot([t - 1, t^2, t=-2..2]);`
Notice the placement of the brackets ([ ]) around the two parametric function rules and the specification of the domain of *t*.

*You can adjust the graphics window using the standard method for specifying the horizontal and vertical axes.*

`plot([t - 1, t^2,t=-2..2],-5..5,-2..10);`
The curve will still appear between $-3$ and $1$ on the horizontal axis, since the parametric rule $t-1$ (the first coordinate) has domain $[-2, 2]$. The axes, however, have values between $[-5, 5]$ horizontally and $[-2, 10]$ vertically. The function rules and the domain for the parameter *t* appear together inside the brackets, followed by the horizontal and vertical specifications for the graphics window. The x and y are not placed on the axes, since they are omitted from the domain and range arguments.

*You can graph more complicated curves using this parametric approach.*

`plot([t - sin(t),1 - cos(t),t=0..2*Pi]);`
The parametric functions are $x(t) = t-\sin(t)$ and $y(t) = 1-\cos(t)$. The domain of the parameter *t* is $[0, 2\pi]$. This curve is called a cycloid, which is easy to describe parametrically but is very complicated to describe in standard function form.

## Polar Form

*The parametric graphing feature allows you to graph functions in polar coordinates.*

`plot([sin(t),t, t=0..Pi],coords=polar);`
This graphs the polar coordinate functions $r = \sin(t)$. You may recall that this function describes a circle. The graph, however, is oval and does not look like a circle.

*You must use a graphics window with the appropriate aspect ratio.*

```
plot([sin(t), t, t=0..Pi], -3..3, -2..2,
coords=polar);
```
The graphics window is 6 units wide and 4 units high. This gives an aspect ratio of 3 to 2, or 1.5. The aspect ratio of your Macintosh screen is about 1.5. Notice that the graph now looks like a circle.

*Some curves are more easily described using polar coordinates.*

```
plot([1 + cos(t), t, t=0..2*Pi], -3..3,
-2..2, coords=polar);
```
This graph is called a cardioid, since it looks like a heart. It is the graph of $r = 1 + \cos(t)$.

*You can adjust the size of the Plot window once the graph is displayed.*

Place the arrow cursor on the Size box in the lower right corner of the Plot window and drag it toward the middle of the screen. If the Size box is not visible, you need to bring it into view by dragging the window up using the Title bar. You should shrink the size of the Plot window until it takes up about one-half the screen. Be sure *not* to click on the Close box in this case. Rather, click on the Session window (region to the right of the graphics region) to return to the standard Maple environment.

*Let's graph another function and display it side by side with the previous function graph.*

```
plot([1 + sin(t), t, t=0..2*Pi], -3..3,
-2..2, coords=polar);
```
Shrink this Plot window to about half the size of the screen. Now use the Title bar to move this Plot window to the right half of the screen. Pull down the Windows menu and select the second menu item on it (the previous plot). You can use the Size boxes and Title bars on these two windows to position them as you wish. Often, comparisons of similar graphs can give you insights into problems.

# 6 Examples

## A Rational Function with Asymptotes

The graphs of rational functions can have interesting features such as horizontal, vertical, and oblique asymptotes. You can investigate these asymptotes by looking at the zeros of the denominators and at the behavior of the functions as the variable approaches $\pm\infty$. You will investigate the following rational function

$$f(x) = \frac{3x^3 - x^2 - 3x + 5}{x^2 - 2x - 1}$$

*First you define the function.*

```
f := proc (x) (3*x^3 - x^2 - 3*x + 5)/
(x^2 - 2*x - 1) end;
```
Notice the parentheses around both the numerator and the denominator.

*You can graph the function to obtain an overview of the behavior of the function.*

```
plot(f);
```
As you recall, Maple uses $[-10, 10]$ as the domain unless you specify the domain. Looking at the graph, there appear to be one zero and two vertical asymptotes.

*You can restrict the domain to obtain more detailed information about the graph.*

```
plot(f, x=-3..2);
```
Both the zero and the vertical asymptotes occur between −3 and 2. Notice that the absolute value of the range values along the $y$-axis have increased. This occurs because the same number of points are calculated for the interval $[-3, 2]$ as for the interval $[-10, 10]$. Thus, there are more points calculated near the asymptotes than before.

*You can change the range as well.*

```
plot(f, x=-3..2, y=-10..50);
```
The behavior of the function between the asymptotes is now more apparent. The seemingly vertical lines are, as you recall, calculated points that are connected. They are not truly asymptotes. You may wish to check this by using the `style = POINT` feature.

*You can determine the exact location of the vertical asymptotes.*

```
solve(denom(f(x))=0);
```
The vertical asymptotes can occur only at the zeros of the denominator. The values displayed are exact, but how would you graph them on the $x$-axis?

*You can find floating-point approximations for these two values.*

```
fsolve(denom(f(x))=0);
```
Although exact values are useful at times, approximate values are also useful.

*You may have noticed that the graph of the function seemed to be a straight line away from the asymptotes.*

```
quo(numer(f(x)), denom(f(x)), x);
```
Check to be sure there are sufficient parentheses that match. The `quo` function returns the polynomial part of the quotient. As you can see, this partial quotient is a linear rule.

*You can check that this is an oblique asymptote.*

```
plot({3*x + 5, f(x)});
```
Can you tell which graph was drawn first? Is the graph of the function close to the line away from the asymptotes? Notice that the function graph intersects the oblique asymptote.

| | |
|---|---|
| *You can adjust both the x and y values to obtain a better graph.* | `plot({3*x + 5, f(x)}, -5..5, -20..20);`<br>This graph gives a clearer picture of the behavior of the function. You may wish to examine the graph more closely to the left of $-5$ by further adjusting the $x$ and $y$ values. You might try $x = -20..0, y = -20..10$. |
| *You can also locate the x- and y-intercepts.* | `solve(f(x)=0);`<br>Once again, it is hard to tell what these possible $x$-intercept values are, but, if you look carefully, two of them are complex. |
| *You can find an approximation of the real root.* | `fsolve(f(x)=0);`<br>Notice that this point is to the left of both asymptotes. |
| *The y-intercept is easy to determine.* | `f(0);`<br>The value of the $y$-intercept occurs when $x = 0$. You can easily make this computation in your head. |
| *You can locate the intersection of the graph of the function and the oblique asymptote.* | `fsolve(3*x + 5 = f(x));`<br>We laughed, too. Do you believe this answer? Can you check this answer using pencil and paper? |

## Finding the Roots of Polynomials

You can use the graphing, factoring, and equation-solving capabilities of Maple to investigate the roots of polynomials.

| | |
|---|---|
| *Let's look at a fifth degree polynomial.* | `p := proc (x) 12*x^5 + 32*x^4 - 57*x^3`<br>`- 213*x^2 - 104*x + 60 end;`<br>This is $12x^5 + 32x^4 - 57x^3 - 213x^2 - 104x + 60$. Again, notice the use of the asterisk to indicate multiplication between coefficients and variables. |

You can display this polynomial function.

`p(x);`
Maple displays the function in standard mathematical
notation.

You can graph this poly-
nomial.

`plot(p);`
The domain is $[-10, 10]$, since none is specified. This
graph gives few details of the behavior of the polynomial
between $-5$ and $5$. The graph of this polynomial, however, does appear to cross the $x$-axis between $-5$ and
$5$.

You can focus your at-
tention on a smaller in-
terval.

`plot(p, -5..5);`
A few more details are discernable from this new graph.
It shows that the polynomial may be zero at several
points between $-5$ and $5$. What is the largest value
of a tick mark on the $y$-axis? You should think about
restricting the $y$ values so that more detail will appear.

You know how to spec-
ify the second coordi-
nate values.

`plot(p, -5..5, -10..10);`
This graph indicates there is one zero to the right of the
origin and three zeros to the left. Further investigation
will show that this plot does not give an accurate picture
of the behavior of the function. Such plots can occur
because Maple uses only 25 points to start its plot of an
expression.

You can try a different
set of $y$ values to investigate further the behav-
ior of the polynomial.

`plot(p, -5..5, -100..100);`
This graph is slightly more rounded and seems to show
the complete behavior of the function between $-5$ and
$5$. How many zeros do there appear to be? What is the
degree of the polynomial?

You can magnify the
graph of the function to
the left of the origin.

`plot(p, -2.5..0);`
Notice that the $y$ values are still between $-100$ and $100$.
Here you can more clearly see that the graph just touches
the $x$-axis at $x = -2$. You might reflect on what this
means.

*You can look at the graph to the right of the origin.*

**plot(p, 0..3);**
The location of the first of the two zeros to the right of the origin can now more easily be approximated.

*You can try to factor this polynomial.*

**factor(p(x));**
Maple has factored this polynomial completely. How many linear factors are there? Can you determine the exact values of the rational zeros of the polynomial?

*You can also use the* solve *function to determine the zeros of the polynomial.*

**solve(p(x)=0);**
Notice the zeros appear in the same order as the factors. How many times does −2 appear? How many times did the factor $(x + 2)$ appear?
   You have used the plot function to investigate the behavior of the polynomial in a situation where all the zeros were rational.

*Let's look at another polynomial.*

**q := proc (x) x^5 + 4*x^2 - 3*x + 5 end;**
Be sure to enter the polynomial function carefully.

*Graph the polynomial.*

**plot(q);**
The graph seems to look the same as the first graph of $p$ just given.

*Graph the polynomial from −5 to 5.*

**plot(q, -5..5);**
This time the polynomial seems to cross the $x$-axis once between −5 and 5.

*You can get a better picture of the graph.*

**plot(q, -5..5, -100..100);**
Now it is clear that the polynomial crosses the $x$-axis only once.

*You can obtain a more detailed graph.*

**plot(q, -5..5, -10..10);**
Much of the graph is cut off above the line $y = 10$. You can, however, more clearly see the behavior of the polynomial near its zero.

Let's factor the polynomial.

`factor(q(x));`
Surprise. Maple is unable to factor this particular polynomial. You know it must have at least one real zero from the Fundamental Theorem of Algebra. Thus, there should be at least one linear factor.

Hope is not lost. You can try the `solve` function.

`solve(q(x)=0);`
The `RootOf` the same polynomial in $z$ indicates that Maple could not find an exact solution.

Let's stay with it. You can try the `fsolve` function.

`fsolve(q(x)=0);`
The approximate value is displayed. It is a negative value that matches the information on the graphs. Without further investigation, you might conclude that the other zeros are complex zeros. You might wish to examine the symbolic form of the polynomial to verify for yourself that there are no other real zeros outside the interval $[-5, 5]$.

The rational zeros of a polynomial can be found using the Maple `solve` or `factor` functions. Irrational and complex zeros are sometimes found by Maple. Polynomials of high degree (5, 8, 10) can be difficult to investigate. The graphing approach is a very useful tool in such situations.

Let's look at one last example.

Enter the polynomial.

`r := proc (x) 2*x^5 + 11*x^4 + 2*x^3 - 51*x^2 - 14*x + 60 end;`
This is $2x^5 + 11x^4 + 2x^3 - 51x^2 - 14x + 60$. Be careful as you enter this polynomial.

You can factor first.

`factor(r(x));`
Maple factors this polynomial into linear and quadratic factors with integer coefficients. Can you determine the nature of the zeros?

You can use the `solve` function.

`solve(r(x)=0);`
All the zeros are displayed. How do they match up with the factors displayed before?

*It might be interesting to graph this polynomial.*

```
plot(r);
```
Does this initial graph give you information that matches the displayed zeros?

The `factor` and `solve` functions may give you all the information you need. When they don't, then it is useful to investigate the polynomial using the `plot` function. You need to be flexible in your use of these tools.

# 7 Calculus

## Calculus I

Maple contains a number of calculus functions or operators. These include differentiation, integration, and limit-taking operators.

### Limits of Functions

*You can take the limit of a function as the variable approaches a fixed number for a defined function f.*

```
f := proc (x) (x^2 - 4) / (x - 2) end;
limit(f(x), x = 2);
```

Remember to press Enter after each semicolon. This is $\lim_{x \to 2} f(x)$, where $f(x)$ is defined to be

$$\frac{x^2 - 4}{x - 2}$$

Notice that $x = 2$ indicates the value that the variable $x$ will approach. The displayed value is the limit. You can see that this function is undefined at $x = 2$ but the limit exists there.

*You can graph the function to check if this limit seems correct.*

```
plot(f(x), x = 1..3);
```

Notice that the graph appears to be a straight line even though you know that the function is undefined at $x = 2$.

Can you draw an accurate graph of this function on the interval $[1, 3]$? Does the limit displayed before appear to be correct?

*You can factor the function as an additional check.*

```
factor(f(x));
```
Can you explain why this factored form is not an accurate replacement of $f(x)$? Look at the definition of $f$ and observe that it is undefined at $x = 2$. Is the factored representation undefined at $x = 2$?

*You can find limits of more complicated functions.*

```
f := proc (x) (x - 4)/(sqrt(x) - 2) end;
limit(f(x), x = 4);
```
The function

$$f(x) = \frac{x - 4}{\sqrt{x} - 2}$$

is not defined at $x = 4$. Does the limit displayed seem correct?

*You can graph the function to check the answer.*

```
plot(f(x), x = 0..5);
```
Once again, the function graph is slightly flawed. Can you draw the correct graph of the function? Does the graph resemble a straight line or is it curved (except at $x = 4$)?

You must remember that the `plot` function in Maple computes only a finite number of points and then draws a smooth curve joining the points. Thus, it can incorrectly represent the graph of a function at or near points of discontinuity where the function is undefined.

*You can determine limits at infinity as well.*

```
g := proc (x) sqrt(x^2 - 4*x) - x end;
limit(g(x), x = infinity);
```
Here you are looking at $\lim_{x \to \infty} g(x)$. You should try rationalizing the numerator using

$$\sqrt{x^2 - 4x} + x$$

to check the result displayed by Maple.

You can use the `plot` function to examine the behavior of the function for large values of $x$.

```
plot(g(x), x = 0..100);
```
The number 100 is not a very large value for $x$, but the graph gives you a feel for the behavior of the function.

You can use much larger values of $x$ to obtain more information about the function.

```
plot(g(x), x = 100..1000);
```
The function seems to be flattening out. What value is the function approaching for values of $x$ near 1000? Does this agree with the value given by Maple and the value you calculated?

You can examine piecewise defined functions at the points where their rules change. At these points, you use Maple's ability to determine left and right limits.

You start by defining the function and graphing it.

```
f := proc (x) if x < 0 then x - 1 else
if x >= 0 then x^2 fi fi end;
plot(f(x), x = -2..2);
```
Notice that an error occurs because the `plot` function is unable to decipher this difficult function definition. Maple can, however, graph this function.

You can use just $f$ in the `plot` function.

```
plot(f, x = -2..2);
```
Now Maple graphs the function. The discontinuity is clearly shown at $x = 0$. Can you use open and closed dots to indicate accurately the behavior of the function at $x = 0$?

You can use one-sided limits to determine $\lim_{x\to0^-} f(x)$ and $\lim_{x\to0^+} f(x)$ for

$$f(x) = \begin{cases} x - 1 & \text{for } x < 0 \\ x^2 & \text{for } x \geq 0 \end{cases}$$

```
limit(x - 1, x = 0, left);
limit(x^2, x = 0, right);
```
Does this function have a limit at $x = 0$? An error message stating `cannot evaluate boolean` occurs if you try `limit(f(x), x, x = 0)`.

## Derivatives of Functions

Maple can differentiate most elementary functions.

*The differentiation operator is* `diff`.

```
diff(3*x^4 - 4*x^2 - 5, x);
```
You should have no problem checking the displayed result. Notice that it is necessary to indicate that you are differentiating with respect to $x$.

*You can differentiate the quotient of functions.*

```
diff((x + 1)^2 / (x^2 + 2*x)^2, x);
```
This is

$$\frac{(x+1)^2}{(x^2+2x)^2}$$

This expression can be simplified.

## Investigating a Function Using `plot` and `diff`

You can investigate the behavior of a function by graphing it. The information you garner from graphs of the function in conjunction with the zeros of its first and second derivatives can give you accurate approximations of the maxima, minima, and points of inflection. Let us consider the function:

$$f(x) = \frac{x+2}{(3+(x^2+1)^3)}$$

*You begin by entering the function as an expression.*

```
f := (x + 2)/(3 + (x^2 + 1)^3);
```
The parentheses are needed in the numerator and denominator to ensure that Maple performs the operations in the order intended. This is a difficult function to investigate using pencil-and-paper methods.

| | |
|---|---|
| *Graph the function.* | **plot(f);**<br>You begin by letting Maple determine the $x$ and $y$ values. Notice that the function seems to appear near $-2$ and seems to disappear near 2 and that the maximum value is less than 1. |
| *You regraph the function with your own values for $x$ and $y$.* | **plot(f, x=-5..5, y=-0.1..0.1);**<br>The $y$ values are indicated as $-0.1$ and $0.1$ to avoid the possible confusion that might occur with multiple dots between the numbers. You can see now that the graph crosses the $x$-axis near $-2$. You should use the Windows menu to select the Maple Sessions window rather than clicking the Close box on the Plot window. |
| *You can determine where maxima, minima, and points of inflection occur by looking at the derivatives of $f(x)$.* | **diff(f, x);**<br>**d := ";**<br>**simplify(d);**<br>**fsolve(numer("));**<br>These four statements give the first derivative and the approximate zeros for the numerator of $f'(x)$. These are, of course, the zeros of $f'(x)$. Referring back to the last graph by selecting the last Plot window from the Windows menu, you can see that the negative number is the position of the minimum for the function. |
| *You can more clearly see the behavior of the function to the left of zero by choosing appropriate $x$ and $y$ values.* | **plot(f, x=-4..0, y=-0.01..0.01);**<br>The zeros before the decimal points are important. Is the minimum more discernible now? Does there appear to be a point of inflection to the left of this minimum? |
| *Now for the points of inflection.* | **diff(d, x);**<br>**simplify(");**<br>**fsolve(numer("));**<br>The expression $d$ is the rule for $f'(x)$. Thus, the expression diff(d, x) is the second derivative of $f(x)$. Once again, you solve the numerator of the second derivative to reduce the amount of work that Maple has to do. This can cause problems if the denominator has zeros. |

Vertical tangent lines and cusps have not been considered here because the graphs did not seem to indicate them. How would you check to make sure there are no vertical tangent lines? Also, you might wish to know about horizontal and vertical asymptotes. You can investigate these using the `limit` function.

You have been working with a very difficult function whose most interesting behavior occurs very close to the $x$-axis. This would be very difficult to discover using pencil-and-paper methods. However, the combination of refining graphs and considering derivatives provides tools to allow you to understand thoroughly the behavior of the function. The analytical skills you can develop investigating such difficult functions with Maple will pay big dividends for you in any quantitative work you do in the future.

### Integrals of Functions

Maple can determine indefinite and definite integrals of functions. Maple also has the ability to approximate definite integrals for functions whose antiderivative cannot be determined.

*Integrals of polynomials are straightforward.*

```
int(3*x^4 - 2*x, x);
```
You must indicate the variable of integration as with differentiation. Can you check the result displayed? *Hint:* Do you remember the `diff` function? Notice that a specific antiderivative is given, not the most general antiderivative. The constant of integration is not displayed.

*Maple can integrate trigonometric functions such as $f(x) = \sec^4(x)$*

```
int(sec(x)^4, x);
```
Notice the placement of the exponent. Can you check the result?

*Maple can perform integration by parts.*

```
int(x^3 * ln(x), x);
```
You can use `diff(", x);` to check the result.

You can integrate more complicated integrals such as

$$\int \frac{x^2}{\sqrt{x^2-9}}dx$$

```
int(x^2 / sqrt(x^2 - 9), x);
```
Be careful when you enter this function.

You can check this result.

```
diff(", x);
```
This does not look anything like the original integrand.

You can, however, simplify this expression.

```
simplify(");
```
Check carefully to make sure this matches the original integrand.

Finally, you can integrate functions such as

$$\frac{x^2+2x+1}{(x^2+1)^2(x-2)}$$

```
int((x^2+2*x+1)/((x^2+1)^2*(x - 2)), x);
```
Clearly, you need to be careful when entering long expressions on one line. Remember, when you check this result, you may need to use the simplify function.

The student package for calculus, available with Maple, allows you to determine the solutions to certain problems in a step-by-step fashion rather than by simply applying a single Maple function. This is very useful when you wish to learn the solution process.

To use the student package for calculus, you must first load it.

```
with(student);
```
This makes the functions in this package available.

You can apply the integration by parts function to appropriate integrals.

```
intparts(Int(x*sin(x),x),x);
```
This is the integral $\int x \sin(x)\,dx$. Notice the capital I in Int. The last expression in the Maple statement is your choice of which factor in the integrand is to be differentiated. You can see that the result contains an integral that is standard and easily integrated.

You can investigate more complicated integrals.

```
intparts(Int(x^2*exp(2*x),x),x^2);
```
This is the integral $\int x^2 e^{2x}\,dx$. The result includes an integral that appears simpler than the original. This should encourage you to apply the integration by parts method again.

*You apply the method again and choose the expression that is to be differentiated.*

`intparts(Int(x*exp(2*x),x),x);`
You should be able to recognize the integral in this result. The factor 2 needs to be inserted in the integrand (with appropriate adjustment) to make it a standard form. You can now finish the problem. You will notice that the signs and multipliers that result from the integration by parts method need to be accounted for. You may find it easier to do this yourself.

## Calculus II

### Series

You can find the value of sums such as $\sum_{i=1}^{5} i^2$, create a Taylor series for a differentiable function, and create power series using recursive formulas.

*You can sum a finite series.*

`sum(k^2, k = 1..5);`
Here the function is the first argument and the range of the index values is second.

*The last value for the index can be a variable.*

`sum(k^2, k = 1..n);`
Notice that this formula is equivalent to the standard one you encountered in precalculus mathematics or in the introduction to integration in your calculus course.

*You can find some infinite sums as well.*

`sum(1/2^k, k = 1..infinity);`
Since this is a geometric series, Maple can determine the sum.

### Taylor Series

A functions that is differentiable can be represented by a infinite series called a Taylor series. The `taylor` function displays a specified number of terms of the Taylor series expanded about a specified value.

*The function $e^x$ is infinitely differentiable.*

```
taylor(exp(x), x = 0, 5);
```
The Taylor series is expanded about $x = 0$, and the first five terms are displayed. Notice the last term displayed. `O(5)` gives an indication of the degree to which the finite polynomial is inexact.

*The Taylor series for more complicated functions can easily be obtained.*

```
taylor(exp(-x)*cos(x), x = 2, 7);
```
Here the Taylor series is expanded about $x = 2$ and seven terms are requested. Notice that the last term displayed is `O(7)`, indicating the degree of accuracy of the approximation. You may wish to graph the function and its approximation as a visual check of how well the function is approximated. You will need to enter the Taylor series (without the order term).

## Functions of Several Variables

Maple can determine the limit of a function of several variables. Maple can also find partial derivatives and multiple integrals of functions of several variables.

### Derivatives of Functions of Several Variables

*You can find the partial derivatives of $f(x, y) = x^2 y^3 + e^x + \ln(y)$.*

```
diff(x^2*y^3 + exp(x) + ln(y), x);
```
This yields the partial derivative of the function with respect to $x$:

$$\frac{\partial f(x, y)}{\partial x}$$

The partial derivative of $f(x,y) = x^2 y^3 + e^x + \ln(y)$ with respect to $y$ can also be computed.

```
diff(x^2*y^3 + exp(x) + ln(y), y);
```
This yields the partial derivative of the function with respect to $y$:

$$\frac{\partial f(x,y)}{\partial y}$$

As you might expect, you can find second partial derivatives.

```
diff(diff(x^2*y^3 + cos(x)*sin(y),x),y);
```
This is

$$\frac{\partial^2 f(x,y)}{\partial x \partial y}$$

Maple can determine other partial derivatives in the same way.

## Multiple Integrals of Functions of Several Variables

You can integrate

$$\int_0^1 \int_0^{\sqrt{1-x}} xy^2 \, dy \, dx$$

```
int(int(x*y^2,y=0..sqrt(1-x)),x = 0..1);
```
The fraction displayed is the exact value of this double integral. Triple integrals follow this same form.

## Limits of a Function of Several Variables

You can take the limit of a function of two (or more) variables as $(x,y)$ approaches $(a,b)$.

```
limit(x*y/(x^2 + y^2), {x = 0, y = 0});
```
This is

$$\lim_{(x,y) \to (0,0)} \frac{xy}{x^2 + y^2}$$

This limit is undefined, since the limit along the $y$-axis is different from the limit along the line $y = x$.

*You can find more complicated limits involving transcendental functions.*

```
limit(x*exp(y - x), {x = 1, y = 0});
```

This is

$$\lim_{(x,y)\to(1,0)} xe^{y-x}$$

The limit is 1/e. Limits of functions of three or more variables are determined in the same way.

## Summary

### Maple Calculus Functions

You can differentiate, integrate, and take limits of functions of one or more variables using the `diff`, `int`, and `limit` functions in Maple. The `simplify` function is often useful in displaying results in a more readable form. The `plot` function can be used to obtain graphic information about a function. This graphic information in conjunction with the first and second derivatives and the `fsolve` function can be useful in investigating the behavior of a function.

You can load the student package for calculus using the `with` function. You can use the `intparts` function in this package to become familiar with the integration by parts technique.

Infinite sums can be investigated using the `sum` function. Taylor series can be obtained using the `taylor` function.

## Additional Activities

You may wish to explore the following functions using the `diff`, `int`, `limit`, `plot`, `simplify`, `fsolve`, `numer`, `denom`, and `taylor` functions.

Accurately locate any maximum and minimum values and points of inflection, and draw a careful sketch of each of the following functions.

1. $f(x) = x^{2/3} - \dfrac{1}{5}x^{5/3}$

2. $f(x) = \dfrac{x^2 + x - 2}{x^2}$

3. $f(x) = \dfrac{\sqrt[3]{1-x}}{1+x^2}$

Evaluate the following limits.

4. $\displaystyle\lim_{x \to \infty} \dfrac{2x^2 - 1}{x^2 + 3}$

5. $\displaystyle\lim_{x \to -1} \dfrac{x^2 - 2x + 1}{2x^2 - x - 3}$

6. $\displaystyle\lim_{(x,y) \to (0,0)} \dfrac{x^3}{x^2 + y^2}$

7. $\displaystyle\lim_{(x,y) \to (1,0)} \dfrac{xy - y}{x^2 + y^2 - 2x + 1}$

Perform the following integrations.

8. $\int x \ln x \, dx$

9. $\displaystyle\int \dfrac{dx}{\sqrt{x^2 - 4}}$

10. $\displaystyle\int \dfrac{2x^2 + 19x - 45}{x^3 - 2x^2 - 5x + 6} \, dx$

**11.** $\int \dfrac{dx}{4\sin x - 3\cos x}$

**12.** $\int e^{-x}\cos x\, dx$

**13.** $\int_0^1 \int_0^{\sqrt{x}} ye^{x^2}\, dy\, dx$

Find and graph the first five terms of the Taylor series expansion about the indicated value $a$ for each of the following.

**14.** $\cos x$,    $a = 0$

**15.** $\sin x$,    $a = 0$

**16.** $\cos x$,    $a = 1$

**17.** $x \sin 2x$,    $a = 0$

For the given functions, show that $\dfrac{\partial^2 f(x, y)}{\partial x \partial y} = \dfrac{\partial^2 f(x, y)}{\partial y \partial x}$.

**18.** $f(x, y) = x^2 y^3 + \cos x \sin y$

**19.** $f(x, y) = 2^x y^2$

# 8  Differential Equations

## Differential Equations

Maple can solve many differential equations as explicit functions (in closed form). When necessary, it can give approximate numerical solutions to such equations. In addition, Maple provides a LaPlace transform capability that can be used in solving differential equations.

*Maple can easily solve the differential equation $y' = y$.*

**dsolve(diff(y(x),x) - y(x) = 0, y(x));**
The differential equation is written in the form $y' - y = 0$, since Maple requires a zero on the right side of the equation. The y(x) at the end of the statement indicates the differential equation is to be solved for the function $y(x)$. Of course, $y(x)$ is the function that is its own derivative ($e^x$). The solution is actually a family of functions based on the constant $C$.

*A series solution is possible.*

**dsolve(diff(y(x),x)-y(x)=0,y(x),series);**
Here the solution is given in series form with six terms and an error indicator, O(7). The number of terms is determined by the value of the variable Order, which is set initially to 6. You can assign a higher or lower numerical value to the variable Order if you wish.

*The LaPlace transform approach can be used if the initial value of $y(0)$ is known.*

**dsolve(diff(y(x),x)-y(x)=0,y(x),laplace);**
This function loads the package called `laplace` and uses it to solve the differential equation. Notice that the solution contains $y(0)$, the initial value of $y$, in place of the constant $C$.

## Initial Value Differential Equations

*More involved differential equations can be solved using the `dsolve` function, for example, $y'' + k^2y = 0$, where $y(0) = 0, y'(0) = k$ are the initial conditions.*

**dsolve(diff(y(x),x,x)+k^2*y(x)=0,y(x), laplace);**
This is a homogeneous differential equation that is solved using LaPlace transforms. The function $y(x)$, which is returned as the solution to this differential equation, contains $y(0)$ and $yp(0)$ (where $yp(0)$ is $y'(0)$).

*You can use the `subs` function to simplify this expression for $y(x)$.*

**subs(y(0) = 0, yp(0) = k, ");**
The displayed solution incorporates the initial conditions in the general solution.

*Maple can also solve non-homogeneous differential equations with initial conditions.*

**dsolve(diff(y(x),x,x)+y(x)-4*exp(x)=0, y(x), laplace);**
This is the differential equation $y''+y-4e^x = 0$ with the initial conditions that the first and second derivatives of $y$ are zero at zero.

*The `subs` function can again be used to incorporate the initial conditions.*

**subs(y(0) = 0, yp(0) = 0, ");**
These substitutions, of course, greatly simplify the result.

## Systems of Linear Differential Equations

Maple can solve systems of equations such as:

$$x''(t) - x(t) + 5y'(t) = t$$
$$y''(t) - 4y(t) - 2x'(t) = -2$$

with initial conditions

$$x(0) = 0, \quad x'(0) = 0, \quad y(0) = 0, \quad y'(0) = 0$$

*The system can be entered and then solved.*

```
sys:=diff(x(t),t,t)-x(t) + 5*diff(y(t),t)
- t = 0,
diff(y(t),t,t) - 4*y(t) - 2*diff(x(t),t)
+ 2=0;
```

```
dsolve({sys,x(0) = 0,xp(0) = 0, y(0) = 0,
yp(0) = 0},{x(t),y(t)},laplace);
```

When a Maple statement will not fit on the screen, press Return (not Enter) between words and finish the statement. You will need to highlight the entire statement before pressing Enter. The system with its initial conditions are enclosed in braces. The equations are entered with zeros as the right sides.

# 9 Linear Algebra

## The linalg Package

The Maple linear algebra package, called linalg contains a number of functions that are useful in the study of matrices and linear transformations and in the solution of systems of linear equations.

*You use the* with *statement to access the* linalg *package.*

**with(linalg);**
The functions that are loaded into the Maple environment are displayed on the screen.

*You enter a matrix with the* array *function.*

**A := array(1..4,1..4,[[1,2,3,4],**
**[2,3,0,-5], [2,-1,1,1],[-2,2,0,-5]]);**
The first two arguments of this function are the row and column dimensions of the array. The third argument of this function is the set of matrix entries. The bracketed sets of numbers represent the row entries of the matrix. Thus, the first row has four entries: $1, 2, 3, 4$. The fourth of the four rows has entries $-2, 2, 0, -5$. The bracketed row entries are nested inside brackets.

*You can attempt to display the matrix.*

**A;**
Notice that A is returned. The array data structure is not displayed, only its name.

*You can, however, display the matrix entries.*

```
print(A);
```
Notice that the dimensions of the matrix along with its row entries are displayed.

*You can add two matrices.*

```
add(A, A);
```
Here you are adding the matrix $A$ to itself. Notice that each entry of the displayed matrix is twice the corresponding entry of the matrix $A$.

*You can interchange two rows.*

```
B := swaprow(A, 3, 4);
```
You may wish to display the matrix $A$ again to compare it with $B$. Notice that this new matrix is assigned to $B$. The swapcol function is used to interchange two columns.

*You can add $A$ and $B$.*

```
add(A, B);
```
The last two rows of this matrix are the same.

*You can save this matrix for later use.*

```
D := ";
```
$D$ is a 4-by-4 matrix whose last two rows are the same.

*You can create a matrix that is a linear combination of $A$ and $B$.*

```
add(A, B, 1, 2);
```
The entries of $A$ are multiplied by 1 added to 2 times the entries of $B$. You can use this extended form of the add function to subtract two matrices.

*You can find the inverse of $A$.*

```
inverse(A);
```
Maple displays rational numbers on a single line with a slash if either the numerator or the denominator has one digit.

*You can check the inverse computation.*

```
multiply(A, ");
```
As you would expect, a 4-by-4 identity matrix with 1s on the main diagonal is displayed. This is one way to create an identity matrix.

*Maple can transpose matrices.*

```
transpose(A);
```
You may wish to display $A$ to check the result. Notice that the rows of $A$ have become the columns of the transposed matrix.

| | |
|---|---|
| *The determinant of a matrix is easily obtained.* | `det(A);`<br>The determinant is a multiple of the denominators of the inverse matrix. You may wish to scroll down the inverse matrix to see that this is so. |
| *You may recall that the determinant of a matrix with two identical rows is zero.* | `print(D);`<br>`det(D);`<br>As you might suspect, the determinant is 0. What should this imply for the inverse of *D*? |
| *You can attempt to compute the inverse of D.* | `inverse(D);`<br>An error message is displayed because the matrix is singular. |

## Basis and Dimension Functions

Maple has a number of functions that are useful in theoretical considerations of a matrix as a linear transformation of the form $T : V \rightarrow V$. The basis functions create bases for specific subspaces of *T*'s domain vector space *V*. The dimension functions determine the dimensions (or rank) of a matrix.

| | |
|---|---|
| *You can find a basis for the kernel or nullspace of a linear transformation.* | `kernel(A);`<br>The {} indicates that the kernel is the zero subspace. |
| *Some kernels are nonzero.* | `kernel(D);`<br>The displayed vector is a basis for the kernel, or nullspace, of *D*. |
| *You can find a basis for the row space of a matrix.* | `rowspace(D);`<br>There are three vectors in this basis, and so the dimension of the row space is 3. |

*You can also find a basis for the column space.*

`colspace(D);`
The number of vectors for this basis is the same as for the row space, but the basis vectors are different. Unfortunately, Maple does not distinguish between row and column vectors. Thus, the basis for the column space appear to be row vectors, although they are not.

*Maple can determine the rank of a matrix.*

`rank(D);`
The rank of a matrix is the number of linearly independent rows (or columns). From your previously displayed results, $rank(D) = 3$.

## Characteristic Values and Vectors

Maple has a set of linear algebra functions that allow you to determine the characteristic values and to determine the characteristic vectors.

*You can enter an interesting 3-by-3 matrix.*

`A := array(1..3,1..3,[[1,1,2],[1,2,1],`
`[2,1,1]]);`
Notice the nesting of the brackets.

*Maple can determine the characteristic or eigenvalues of a matrix.*

`eigenvals(A);`
Notice that there are three characteristic values. These are the solutions to the characteristic polynomial.

*You will need to create a 3-by-3 identity matrix.*

`Id := multiply(A,inverse(A));`
The displayed matrix is a 3-by-3 matrix with 1s on the main diagonal and 0s elsewhere.

To determine the characteristic vectors from these characteristic values, you need to recall that a characteristic value *lambda* is a number such that for some vector $x$, $Ax = \lambda x$. From your previous computations, you know that one of the characteristic values is 4. Thus, you are looking for a vector that satisfies the matrix equation

$Ax = 4x$. This can be rewritten as $Ax - 4Ix = 0$, or $(A - 4I)x = 0$, where $I$ is the identity matrix.

| | |
|---|---|
| *You create the matrix* $(A - 4I)$. | **Aminus4I := add(A, Id, 1, -4);**<br>This creates the required matrix, since Id is the 3-by-3 identity matrix. |
| *Maple can solve the matrix equation* $(A - 4I)x = 0$ | **linsolve(Aminus4I,array(1..3,[0, 0, 0]));**<br>The function linsolve has two arguments. One is the array $(A - 4I)$ and the other is the vector $(0, 0, 0)$. The displayed result is a parametrized set of vectors. |
| *You can check to see if this set of vectors satisfies the equation* $Ax = 4x$. | **multiply(A,");**<br>This is the product of the matrix $A$ and the characteristic vector. |

You might want to find and check the other two characteristic vectors.

# Index